The Tasmanian DEVIL

Eric Guiler

St. David's Park Publishing,
GPO Box 307C, Hobart, Australia 7001
1992

National Library of Australia
Cataloguing-in-Publication Data

Guiler, E. R. (Eric Rowland), 1922-
The Tasmanian Devil

ISBN 0 7246 2257 8.

1. Tasmanian devil. I. Title

599.2

Note re illustrations:

All illustrations, photographs and diagrams
are attributed to the author unless otherwise indicated.

Design and layout: Desk top Publishing Department
Tasmanian Government Printing Office

Printed and bound by Tasmanian Government Printing Office,
on fine paper produced in Tasmania.

1: Introduction

This book with pictures tells the story of the life of the intriguing Tasmanian devil although it does not attempt to cover the intriguing and complicated aspects of the devil's physiology and biochemistry which are outside the scope of this brief review.

In recent years the Tasmanian devil has become well-known to many people, particularly in the United States of America, through the publication of a comic strip which portrays the animal as a fearsome creature with great powers of destruction, strength and athletic ability but the real Tasmanian devil is far from being like this popular charismatic animal. No doubt some of the admirers of this comic strip character have a vague idea of where Tasmania lies but very few would know anything of the life and habits of the wild Tasmanian devil.

When the first European settlers encountered a fierce noisy creature unlike any other animal they had ever seen, which had a habit of growling, yelling and screaming when annoyed, they named it the Tasmania devil.

It is an unattractive creature; indeed it is ugly, unfriendly and smells objectionably strong, particularly the male. It is strictly nocturnal and even in areas where the animals are abundant they are seldom seen abroad in daylight. They avoid human contact and can be dangerous to handle. But the very young, like the young of most mammals, are cute and playful.

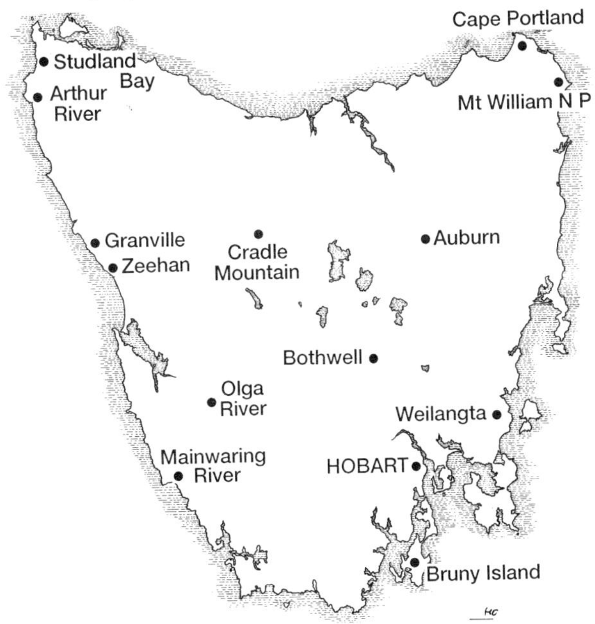

Outline map of Tasmania showing the places mentioned in the text

For all their spectacular displays of teeth and aggression devils are not good fighters and the average farm dog can cope easily with a belligerent devil. Many years ago at Cape Portland (see map on this page) an Australian kelpie sheep dog had a litter of pups in her kennel and the devils sensed an easy meal. By the next morning the pups and mother were still there and six dead devils were evidence of a hard night's work by the mother.

Devils were first seen by Europeans around Hobart when settlers moved in at the beginning of the nineteenth century; devils are still found in the area. They were numerous elsewhere; one record from the east coast claimed that 143 were killed in the winter of 1825 and a large grazing property in the North-west offered a reward for devils in the 1830s.

Tasmanian devil populations suffer collapses from time to time and may not recover for many years. The cause of these crashes is not known. Devils were scarce around 1850 but had built up again by the 1890s. Then, after the turn of the century, the population crashed again and by 1909 devils were found only in the remote districts of the island. Other marsupials suffer also from these crashes and in 1909 all of the family of marsupials to which the devil belongs (Dasyuridae) were affected.

In 1950 a Tasmanian devil was brought into the Zoology Department at the University of Tasmania and everyone rushed to see this unusual animal. By 1955 devils were being found in various scattered localities and then quite suddenly by the 1960s they were back in their old haunts and since then the population has increased to many thousands, if not hundreds of thousands, throughout Tasmania.

Devils are scavengers and feed mainly on dead animals and so are found mainly where game is plentiful. They have two new sources of food in dead domestic stock and garbage tips which have sprung up in and around towns and farms. In this respect the ecology of the Tasmanian devil has changed very considerably over the past two centuries. The devil has profited by the agricultural development of the country which is in marked contrast to the fate of so many other species of mammals throughout the world.

It is recorded that the Tasmanian Aborigines ate devil meat and it is alleged that in early colonial times the convicts also did so. It was said to taste like veal but this is largely hearsay and perhaps not too accurate. Certainly the meat has not been eaten in modern times. The fur is totally valueless so the devil has not been hunted for either its meat or fur since European settlement began 200 years or so ago and this has been much to the species' advantage.

Some Tasmanian animal parks feature displays of devils and these have proved popular and instructive but the animals require a lot of care and attention to keep them clean and not smelling too offensively. Several overseas zoos have kept devils at times but they are not much of a success as exhibits since the animals tend to lie around sleeping all day.

A black devil warily watches our approach.

2: Tasmania - The Devil's Home

The island of Tasmania lies off the south-east corner of Australia. Originally it was connected to the continent but about 90 million or so years ago it was separated from the mainland by continental drift. The dividing sea is shallow and, from time to time, the two land masses have been joined again with a fall in sea levels. This happened last about 13,000 years ago.

Fossil remains of devils are found all over the mainland from Cape York in the north-east to Western Australia and Victoria. Some have been dated as being only 3000 years old. An intriguing discovery of devil bones at Lake Coorangamite, Victoria, was made in 1912 where the bones were unmineralised and apparently fresh. A live devil turned up at Tooborac, one hundred kilometres from Melbourne in 1912 and another at Ballarat in 1971. It is not known whether these animals were from the wild or had been brought from Tasmania and then escaped or had been released.

Tasmanian devils no longer live on any of the Tasmanian offshore islands, although sub-fossil bones have been found on Flinders Island. The Bruny Island and Southern tribes of Aborigines used the onomatopoeic word 'tarrabah' for the devil which is a fair representation of its aggressive growl.

Tasmania is about 68000 square kilometres in area roughly the same size as Ireland but smaller than West Virginia or Honshu. A considerable part of the country is mountainous although it does not often rise above 1000 metres. There are no permanent snowfields.

The island is sparsely populated with very few large towns of more than 5000 inhabitants. Nearly all of the south-west part of the island is a World Heritage Area with few roads and access is mainly by boat or walking track.

1. In valleys such as this one in Cradle Mountain National Park the vegetation is so dense that it is difficult for large mammals to move through it and there is little ground cover or vegetation suitable for food for wallabies. Consequently few devils are here and the rocky mountain tops also offer no feeding and similarly support few animals.

2 Much of Tasmania is covered by eucalypt forests which have high trees through which animals can move freely, often making trails. This type of country offers shelter, food and den sites and has dense populations of mammals including devils. These forests range from wet areas of 150cm or more annual rainfall in the west to dry areas of 75cm rainfall on the east coast. Weilangta State Forest.

3 The very dense undergrowth and absence of suitable grazing makes this rainforest habitat unsuitable for larger mammals and few devils are found here. This country can have more than 250mm annual rainfall. Rainforest, Arthur River.

4 The extensive sedgelands, mainly found in western districts, offer little shelter for larger mammals and few are found here. The sedge, known as buttongrass, is not palatable and the ground is usually wet and boggy. Sedgelands, south of Zeehan.

5 This is ideal game country having excellent shelter, good grazing and a mild climate. Many devils are found in this type of country. Coastal scrub, Granville Harbour.

6 An adult male devil going about its business. The devil has a bulky body with strong forelimbs, massive head and weak-looking back legs. It is about the same size as a spaniel but has few of the dog's endearing characteristics. Cape Portland.

7 The male is larger than the female and has a broader head, more powerful build and carries many more facial scars.

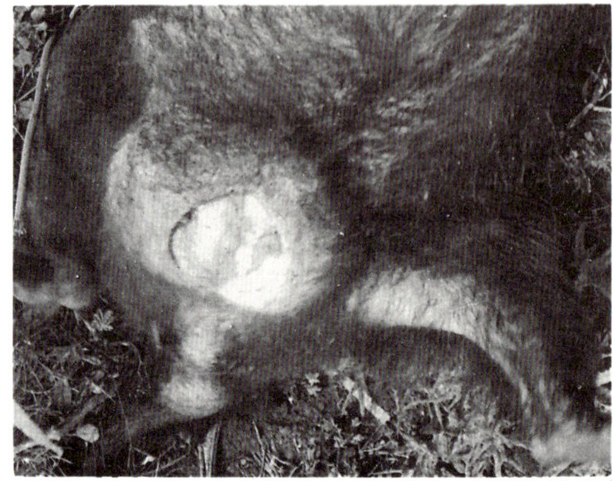

9 The female has a posterior abdominal pouch for the young; the entrance faces backwards to protect the young when the mother is moving through the bush. The pouch has four nipples. This pouch is distended by swollen milk glands and obviously she has been separated from her young for several hours.

8 The general body colour of devils is black relieved by one or more white patches which are useful in identifying individuals. Some devils are entirely black. This photograph of a devil from Bothwell taken in 1974 is the only known record of a white animal and it was in poor condition.

3: Summer December-February

The devil's year has been divided into four three-month periods, roughly corresponding to the four seasons. It is appropriate to start with summer as this is the season of abundance, warmth and easy living. It is also the time when the population of devils running around the bush is greatest.

During summer the climate is mild and food is plentiful. Young wallabies and other marsupials have left their mothers and now are finding their way about but many of these will die before they have learned to forage efficiently for themselves. In addition to this supply of food an agile devil can catch insects, especially grubs and grasshoppers, and lizards as well as small snakes.

At this time of year the devil population consists of adults, year-old immatures and the very young who have just left or are leaving their mothers. This last group are learning to cope with their environment and they, like the other very young marsupials, have a high mortality rate. The greatest number of deaths occurs during the first two months of independent life. However, many of the devils up to two years old do not survive.

Although many hundreds of devils die each year it is a very rare event to find a devil carcase in the bush. Devils are cannibals and eat their own kind, making a good clean job of it.

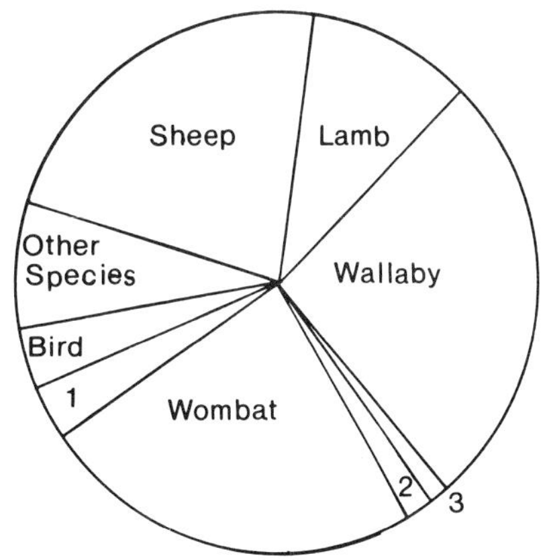

10 Devils are attracted into areas where there is an abundance of food so their numbers can be expected to vary dramatically from time to time. This occurs frequently nowadays with the large number of deaths among domestic animals especially during calving and lambing time and at garbage tips. This diagram shows the food eaten by devils at Cape Portland. The importance of domestic animals and wallabies is illustrated. In other places cattle could be of more importance. Sector 1 indicates plants other than grass eaten; 2 indicates grass; 3 indicates unidentified material.

Sheep meat is eaten mostly as carrion. Unless a sheep is weak and unable to defend itself it is safe from predation. However, a sheep that is lambing is vulnerable as is the newly-born lamb. It is recorded that a devil ate the udder off a cow which was unable to stand up.

Devils pick up scraps as they wander along and will gorge to repletion if they find a good source of food, an adult male being able to eat up to a quarter of its body weight at one sitting. A gastronomic feat at Granville was the devouring of a 16-hand horse when after three days only some bones and the tail hairs were left.

Stomach contents, unless they contain maggots, do not show whether the food was eaten as carrion or killed and then eaten. Laboratory experiments suggest that large adult males are poor predators and field observations are scarce and unreliable. However, it is recorded that a young devil dug out a rabbit burrow and ate the young.

Devils do not eat vegetable matter, any grass found in their stomachs probably has been ingested accidentally but otherwise they will eat anything that is digestible or looks as if it is. They will return to a cow skeleton for months after it was killed. They will eat rubber boots and plastic containers. Devils living near the seashore spend time beachcombing and will eat fish, sea-squirts and anything else but they are not known to attack stranded whales or their carcases. There are limits, even for a devil.

There are no records of devils attacking sleeping humans but they will eat corpses.

12 Feeding is accompanied by much squabbling, loud screams, growls, jaw chomping, jostling and general aggression. This is particularly true of males which often waste much feeding time. Not much serious damage is inflicted except for nips and bites, much of the aggression being a ritualistic display. Devils do not gyrate rapidly like their comic strip character. Drawing: O. L. K. Buckman.

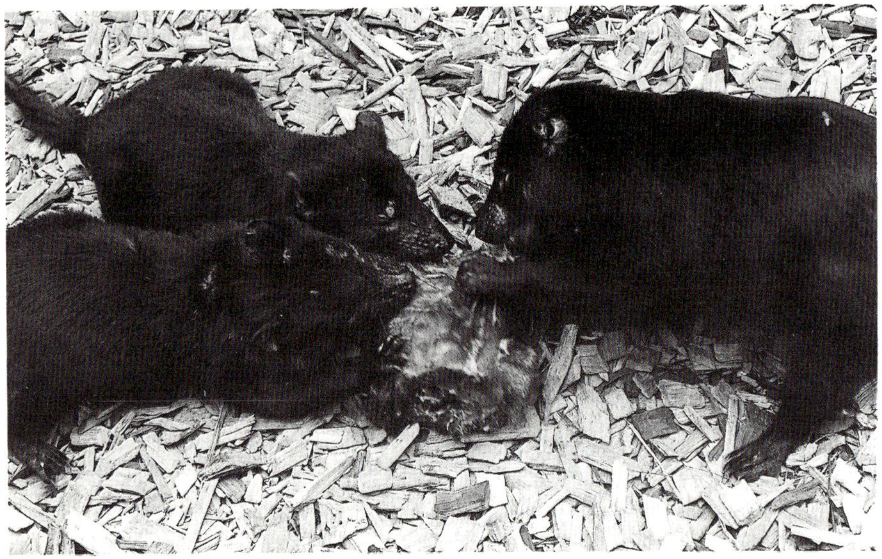

11 Devils congregate at food sites. These three were photographed at the Tasmanian Devil Park eating a piece of kangaroo. The devil on the right is using its forepaw to hold and manipulate the food. This is common behaviour. As many as seven devils have been seen feeding on a dead calf at Cape Portland.

13 This battered veteran from Cape Portland had pieces of the lips, jowls and face torn away. Females do not show as much damage, probably keeping away from the cantankerous males.

At times serious combats take place and males show extensive facial scars. One male discovered during our researches at Cape Portland had no skin over the whole rump exposing the raw flesh. The wound was clean and the animal in good condition and we were horrified to see it run away and dive into a gorse bush for cover.

Internal damage to the facial structures also occurs but devils appear to have considerable recuperative powers. Skulls collected from poisoned animals sometimes show healing of severe injuries. One male had both frontal bones shattered over the brain leaving a hole which extended into the cranial cavity. This was probably the result of a severe blow to the head. The hole was almost fully repaired leaving only a small aperture 1.6mm in diameter.

Another devil's skull showed a repaired wound from a .22 bullet. There was no exit hole so the devil must have carried the bullet around in its head.

Other repaired bones noted during our research were a shattered zygomatic arch and various depressed fractures to the facial and snout bones. The zygomatic arch is made up of the strong bones lying below each orbit and forming the anchorage for some of the powerful jaw muscles.

In addition, tooth damage was almost universal particularly in the older animals and tooth wear of the molars was extensive.

In the light of the above evidence it is obvious that devils are very tough and a small amount of facial skin damage would be of little inconvenience.

14 Young up to about a year old can climb scrubby low bushes and small trees up to about ten metres. Photo: Al Milotte.

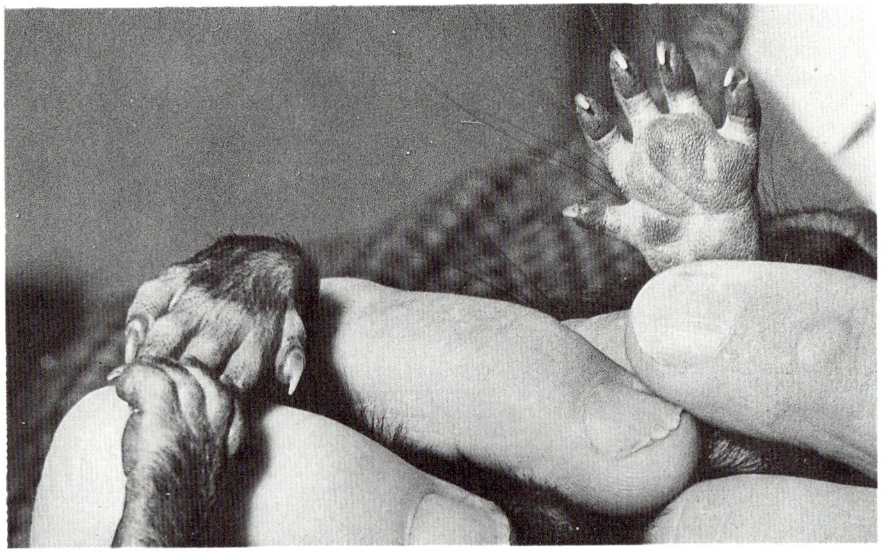

15 The young have mobile forepaws and wrists which assist them in climbing. Getting down from the tree is easy - they just jump or fall. This climbing ability offers the chance of catching birds roosting at night. It is not unusual to find feathers in a devil's stomach.

16 Devils which, as mentioned, are totally nocturnal, spend the day sleeping in a den. This may be in burrows in sand dunes, log piles, hollow logs, caves, buildings, hollows under thick vegetation or indeed any handy spot for sleeping off a feed. This burrow at Cape Portland was shared with a wombat. Many dens are hard to find and neither grass nor vegetation is carried in to make a bed.

More than one devil will use the same den and a devil may have several dens, using each for a few days as needed. Devils do not move far from a food supply to rest. We found three devils sleeping off their feast inside the rib cage of a cow which they were consuming.

A den is used also for mating. The female leaves her young in a den when she goes out foraging but she does not share this with a male as he would probably eat the young. Male devils have no qualms about such behaviour.

Apart from when they are in a den, devils lead a solitary life, never moving in pairs or groups except when the young are running with the mother. They have not been shown to form packs to hunt down prey, although some farmers content that they do this to kill sheep. Apart from mating their only social contacts are at a food source when they behave in a most anti-social way with their squabbling.

Adult devils' main predators are dogs and man. Very young devils would probably be taken by hawks, feral cats, dogs, tiger cats and adult devils. Some female devils under duress will eat their own young ones.

Constant roving in search of food, a largely solitary social life and constant squabbling over food with gargantuan feeds whenever possible is the daily pattern of life of devils throughout the year with little friendly contact with any animal except perhaps a shared den for a short while during the day.

4: Autumn, March-May

By the end of summer food is still plentiful but the good times are drawing to a close. The days are still warm but the nights are cooler, especially at higher altitudes.

The hours of darkness are lengthening giving the devils more time for scavenging as food becomes harder to find. The females have had a time of recuperation from the nursing of, and caring for, their young and the year-old devils are establishing themselves in the community.

Adult females weigh about 6 kilograms and the males are heavier at around 8 kilograms but heavier animals of both sexes are common.

Mating begins at the end of February and continues through to March. The young are born in April after a gestation period of 18-21 days and the females with young will be kept busy right through to December looking after them. Young have been found in the pouch late in the year so some out-of-phase breeding must occur occasionally.

17 Young devils, about one year old, have soft fur and their markings are clear and distinct. The absence of facial scarring makes them quite attractive. They are plump, weighing about 3 kilograms.

Mating in a largely solitary quarrelsome animal requires some temporary psychological adjustment but, apart from this, the male's routine does not change and he plays no part in the rearing of the young.

The female breeds at two years of age although two instances of one year olds having young in the pouch have been recorded. The male also probably breeds in his second year although they may be physiologically capable of breeding earlier than this. The partners remain together for several days during mating, perhaps even as long as a week.

Mating takes place in the den and lasts over four to eight hours. The female sheds a large number of eggs (114 have been reported) but only about 20 of these develop in the uterus. The growth of the embryos is rapid.

The female remains standing during the birth. Nearly all of the 20 eggs which have developed in the uterus have become embryos. All of these are born but only four can be accommodated in the pouch since there are only four nipples. Probably it is a case of first in, first seated. The rest of the embryos are destined to die. This over-production of young helps ensure that sufficient of them survive the journey from the uterus to the outside world and then along the belly fur to the pouch. During this hazardous journey they are not assisted by the mother, other than by some licking of the fur. Disturbance to the mother, aggressive encounters with other animals or maternal inattention reduce the chances of the young making it to the pouch.

As with all marsupials the young devil is born at a very early stage of development, measuring less than 1cm, at birth. It moves quickly to the pouch and once there it searches around for a nipple. This is achieved by touch only as the young is totally blind at this stage. It is truly remarkable that such a small, undeveloped creature can undertake this journey and have the energy to search around in the pouch. The young attaches itself to the nipple which then swells in its mouth so that the young cannot be removed without being damaged. This ensures that the young do not fall out of the pouch and is an important survival factor.

About 75% of females of breeding age carry young. The two year old females consistently have the largest litter size, averaging as high as 3.59 young per animal; this figure decreases for older animals.

By May the hind limb buds are becoming more developed and stronger, the sex can be determined and the eyes are pigmented with the eyeslits becoming differentiated although the eyes are not yet fully formed or open.

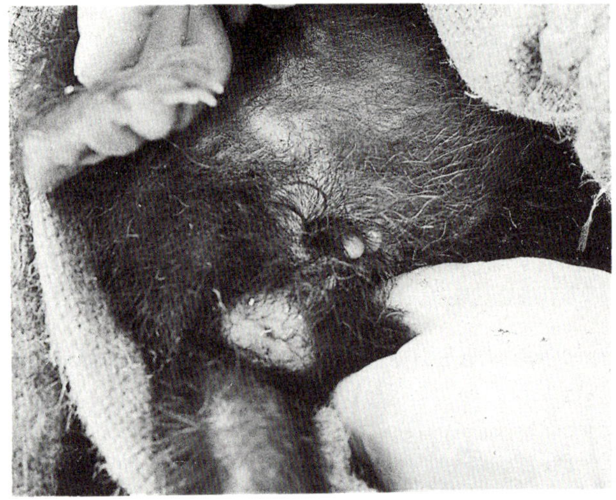

18 Two pouch young at about 7-10 days old. At this stage they are firmly attached to the nipple from which they obtain their milk supply.

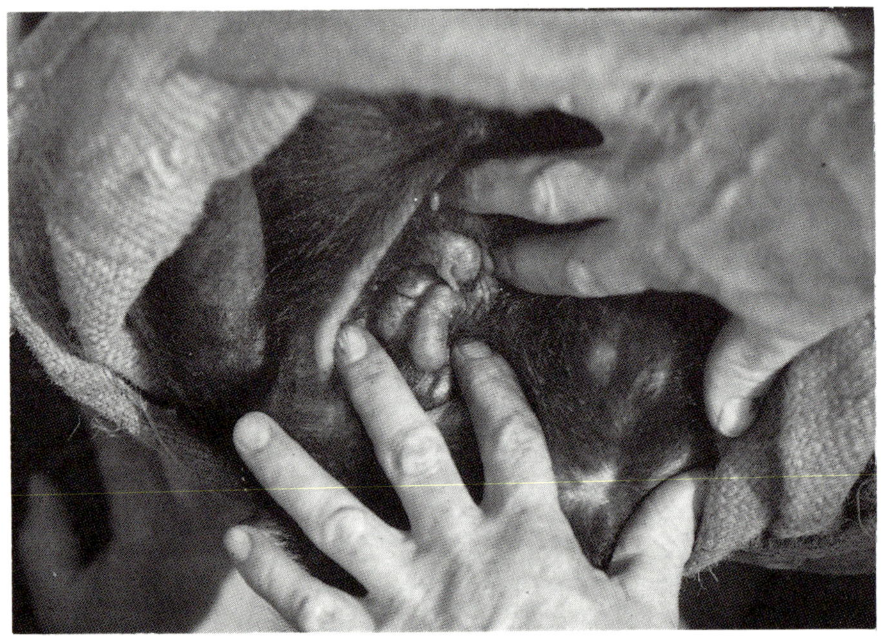

19 Three young at about 20 days old. The head is still proportionally large but the ears are appearing on the sides of the head, although for sometime they will still be applied to the sides of the head. The rows of whiskers on the face can be distinguished and the lips are forming. The eyelashes have not yet appeared but can be seen as a dark pigmented line on the lids. Devils of this size are becoming easier to handle because they have lost the slipperiness of the very young.

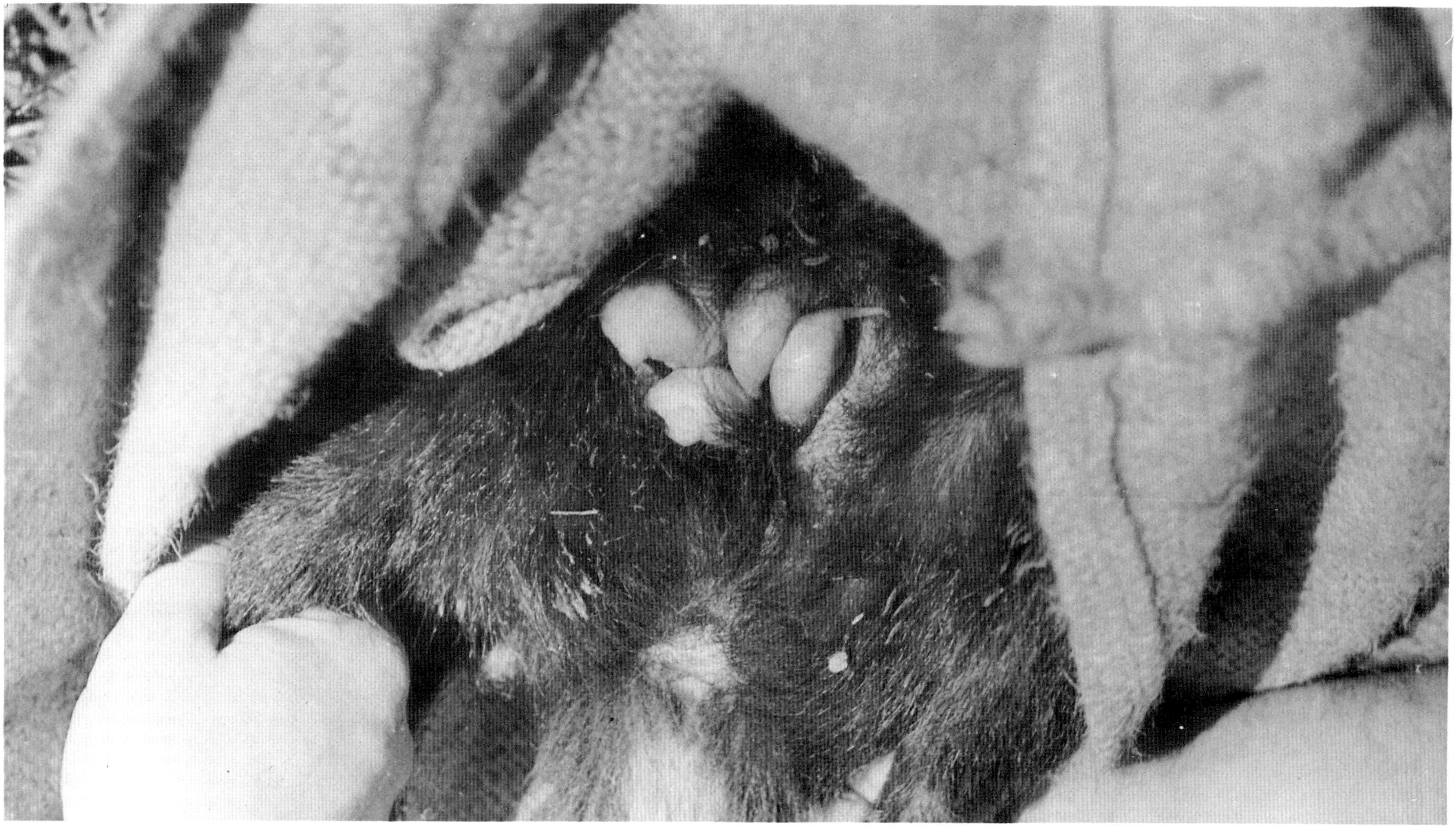

20 A full pouch of four young. The predominantly pink colour is lost as the skin thickens and soon the body will be a dull white colour. Towards the end of May the pigmentation of the body will commence. This process starts at the tip of the nose and gradually extends backwards. In May the young start making squeaking noises although they are still firmly attached to the nipple.

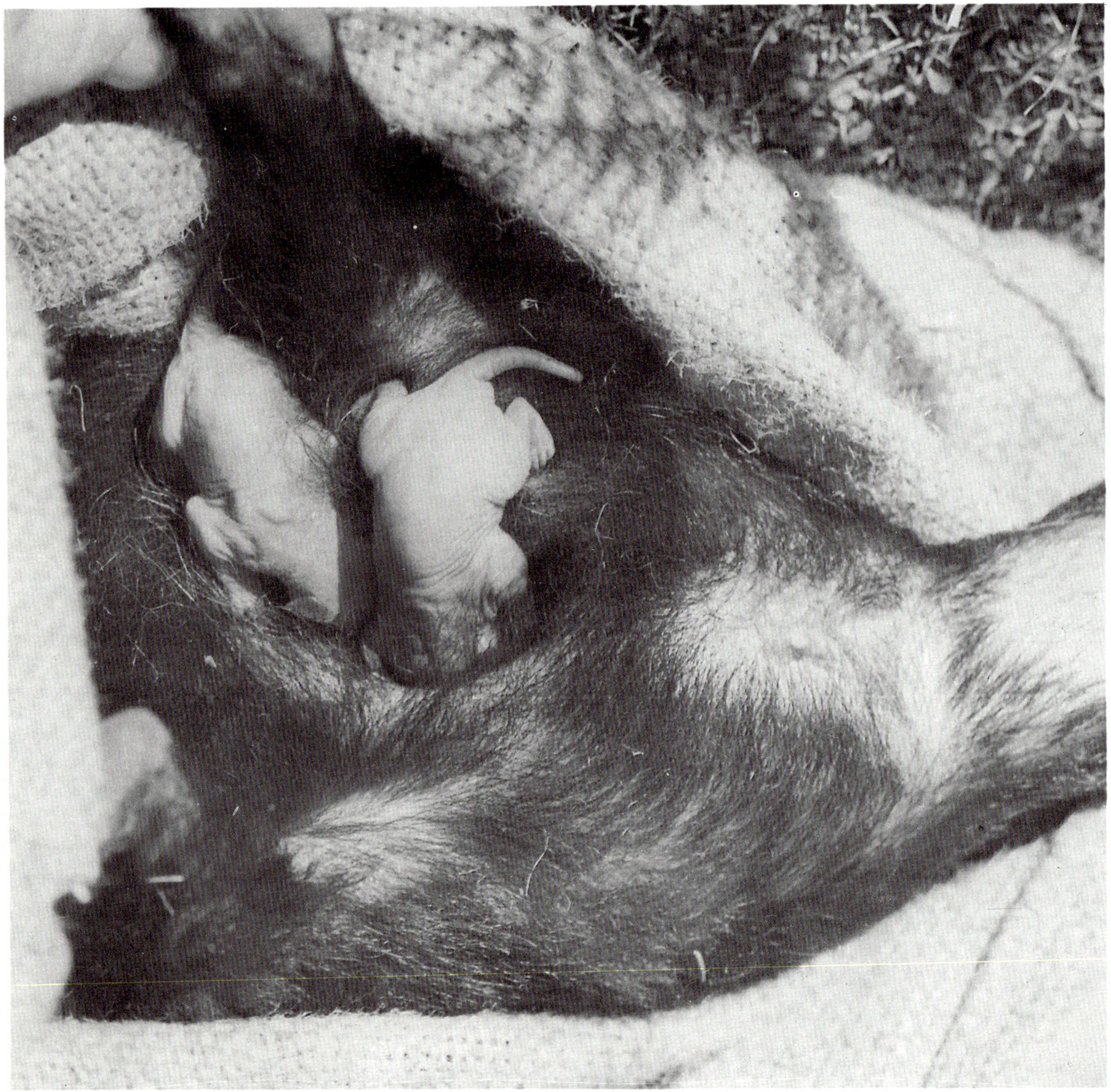

21 These two pouch young are grimly holding on to the teat and using their forepaws to grasp their mother's fur.

5: Winter. June-August

This is the time when the young begin to lose their embryo-like appearance and by the end of June the more advanced young start to grow fur. This follows the same pattern as the body pigmentation, starting at the snout and gradually extending to the rump. The general body colour pattern of the adult becomes apparent, the future white colour patches appearing as non-pigmented areas on the skin.

No deaths of pouch young were recorded among 97 females at Granville during our research, although the young were frequently handled during their pouch life.

23 The furring is virtually complete by the end of the winter. The completion of furring varies by about a week within one litter.

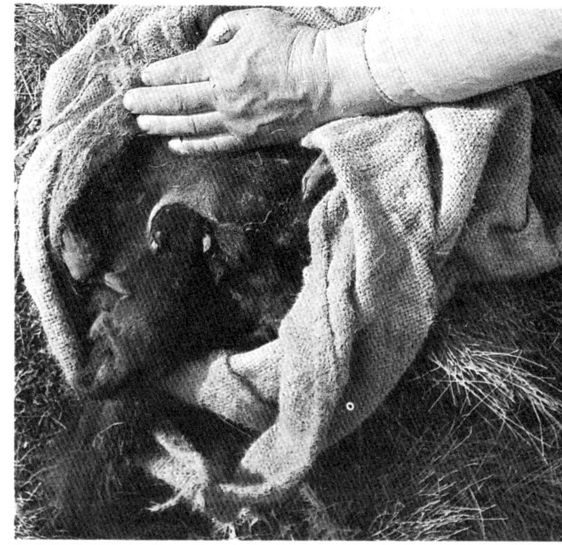

24 By late winter the pouch is becoming crowded. The posterior belly is distended by wriggling young and one, or part thereof, may be left hanging out.

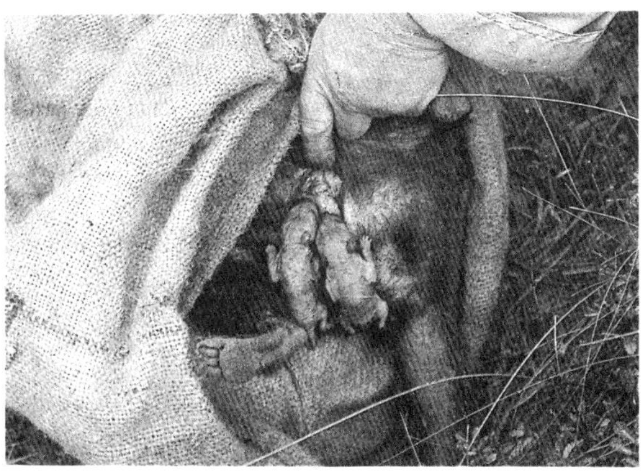

22 The young at about $2\frac{1}{2}$ months of age have reached the stage where they can easily be recognised as young devils. The mouth can be opened and the grip on the teat relaxed. More importantly, the fur is beginning to appear, growing rapidly from the snout down over the rest of the body and by the end of July it reaches the rump. Female devils are very successful in rearing their young at least to the stage when they leave the pouch.

Throughout the last century and up until fairly recently open seasons were held for some of the fur-bearing marsupials such as the wallabies and possums. These were held during the winter months when pelts were in peak condition and devils soon learned that they could find good eating along the snare lines. Trappers resorted to setting traps around their snares to catch the marauders so that in the end this source of food became a death trap. The fur of the devils was useless and the carcases were dumped in the bush where other devils would eat them. But devils have had and continue to have a useful role to play for man. A devil in good condition has a lot of subcutaneous and body fat and in the early days of white settlement its fat was used as a preservative for leather. Devils help control some stock diseases, especially in sheep, by eating up carcases. The devil is not an intermediate host in the life cycle of sheep tapeworms so when an infected carcase is eaten the cycle is broken and the parasite is not passed on. Blowflies are a great scourage throughout Australia, their larvae developing in animal matter such as wounds, the wool of sheep and in carcases. By eating dead animals, with or without maggots, devils help reduce slightly the blowfly population.

26 Devil droppings can be seen along trails in the bush and are sometimes scattered about a small area, the so-called latrines. The droppings are distinctive containing hair and bone fragments, sometimes quite large even as big as a rabbit's foot. Examination of the droppings gives an indication of the food the devil has been eating.

25 August is the beginning of trials for the pouch young. They have grown so much that there is little room for them in the pouch and they have to hold on tightly to the nipple or their mother's fur. Late winter or early spring sees the young left in a den while the mother forages and brings back pieces of food for them. During this time they are alone and virtually defenceless and liable to be eaten by any wandering devil, tiger cat or other predator. The most severe weather of winter is in July-August, particularly in the highlands which have heavy rain and often snow on the peaks. This is a time of hardship for most native animals as food is scarce. Devils may fare better than most other species since, being carrion eaters, they devour other animals which have perished and in doing this they clean up the bush. Dead animals, including devils, are seldom seen except on roads where they have been run over by vehicles.

27 Devils frequently make their droppings along the middle of a track and in time these form mossy patches which eventually become the basis for plants.

28 Ouch. A claw possibly from a bird, in a devil dropping.

Devils forage over a home range but are not territorial guarding their area from intruders. Each devil has its own range which may vary from time to time depending on the availability of food, mating, a population pressures and other factors. The areas of the ranges for males and females appear to be similar with much overlapping of the ranges. As many as ten different animals could be caught in the same trap over a short period of trapping. Devils caught in any specific area appear to fall into two categories – those who live in and around the area and those who appear to pass through or perhaps may not have a greatly extended home range. The resident animals travel up to 16 kilometres in a night seeking food. One non-resident male at Cape Portland was recaptured 25 kilometers away. We experimented with radio tracking but this was not successful as the deep gullies and dense vegetation interfered with the signal.

30 It seems appropriate to end the winter with a youngster a cute little handful, looking forward to spring and its life ahead.

29 A devil is not a rapid mover. It lopes along at about 3-4 kilometres per hour, but when chased it can make about 12 kilometres per hour for a short time. Several times we have caught devils by running after them when they have escaped while being handled.

6: Spring. September – November

In early spring the young start the process of learning to fend for themselves. They are still dependent upon the mother but there is room no longer for them in the pouch so they are left in a den while she forages.

Travelling on the mother's back is a great hazard for young devils as they may fall off or be knocked off by branches. The mother and her young would certainly be the object of aggression at a feeding site and the young would be vulnerable in this situation.

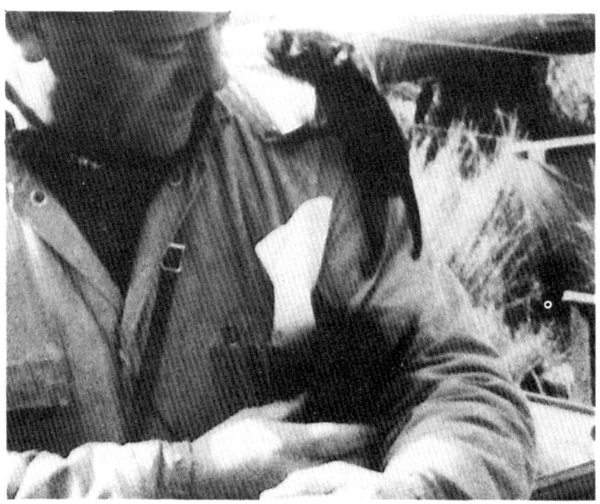

32 In October the young are quite unafraid and are easy to handle, apart from a strong tendency to wriggle.

31 The young were seen being carried around on the mother's back on only one occasion. They held on by their unusually strong forelimbs. We were lucky at Cape Portland to trap a female with attached young. This was the only time anybody has observed this behaviour. I sat down beside the trap and to my astonishment the young devils abandoned their mother and climbed over me to my head where they firmly grasped my beanie and hair and were very difficult to remove.

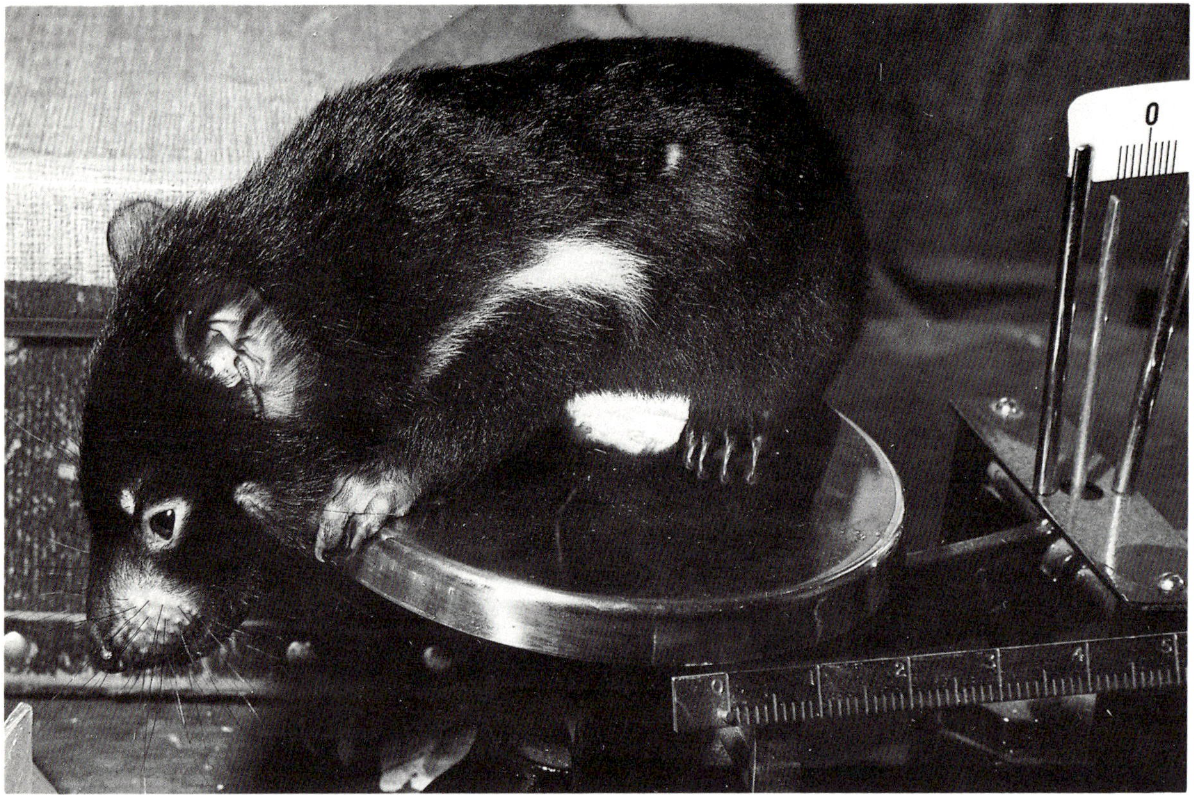

33 What's this all about? They sit still long enough to be weighed.

35 A melanic (all black) devil found at Granville.

When the young start to lead a separate life in November although still cute (weighing about 0.6 kilograms) they have to be handled with care because they can bite like a pair of pliers. Young devils of between seven and ten months of age were not attracted by the baits we used in our cage traps. This suggests they have a different diet from the adults.

36 This is the same little fellow as in the previous picture. He has a firm hold on my pen and thinks he is doing a good job defending himself.

37 In spring the young of the previous year are approaching adulthood ready for the next breeding season in autumn. They are plump and agile and are difficult to handle. October and November are good months for field studies. Food is not yet abundant so there is plenty of stimulus for devils to enter traps in search of meals. The population composition is changing and monitoring reveals the survivors from the winter and the new arrivals into the community.

38 Any dead animals are devoured very rapidly by devils. This cow at Granville died only three days before this photo was taken. The surrounding ground was churned into liquid mud by the trampling of many devils. A large male devil was found at Mount William National Park lying badly injured after being run over by a car. We placed him under a bush to give him a chance of survival, by the next morning he had been eaten, all that remained was the skull. He had fought hard to survive as there was blood splattered everywhere.

7: Catching Devils

By the early 1960 there were as mentioned large numbers of devils in many parts of Tasmania. Nothing was known of their habits and we decided to find out some basic facts about the animal and how it lived. Two main study areas were selected, Cape Portland in the north-east and Granville on the west coast. Both were grazing properties beside the coast and a long way from our base at Hobart. Most work was carried out at Granville as it was a less highly developed area surrounded by forest offering food and shelter for animals. It was more isolated and consequently there was less human interference and the animals had a more natural existence. We trapped devils for several weeks at a time during all seasons for thirteen years and built up a picture of their activities.

40 Our next trap was more successful and the basic design is still in use today. Galvanised reinforcing mesh was used with galvanised sheet iron at one end and a sliding trap door at the other. The door was operated by a simple lever which activated when the baited hook was disturbed. The traps weighed about 8 kilograms and four could be carried by one person.

39 Our first devil trap. This was useless because the devil chewed its way through the cyclone wire mesh at the end of the trap. However, it showed us that devils would enter a trap to eat the beef bait. This type of wire would be strong enough to restrain dogs or even pigs but the devil's jaws are so strong they can tear the mesh apart. The trap must have a bait that will attract the target animal, then catch it and hold it until it can be examined. The trap should not damage the animal or subject it to such trauma that it will never enter a trap again.

41 Traps were often damaged by devils trying to escape. This trap with the unsuccessful galvanised sheet iron floor had already had one side replaced and needed its roof repairing.

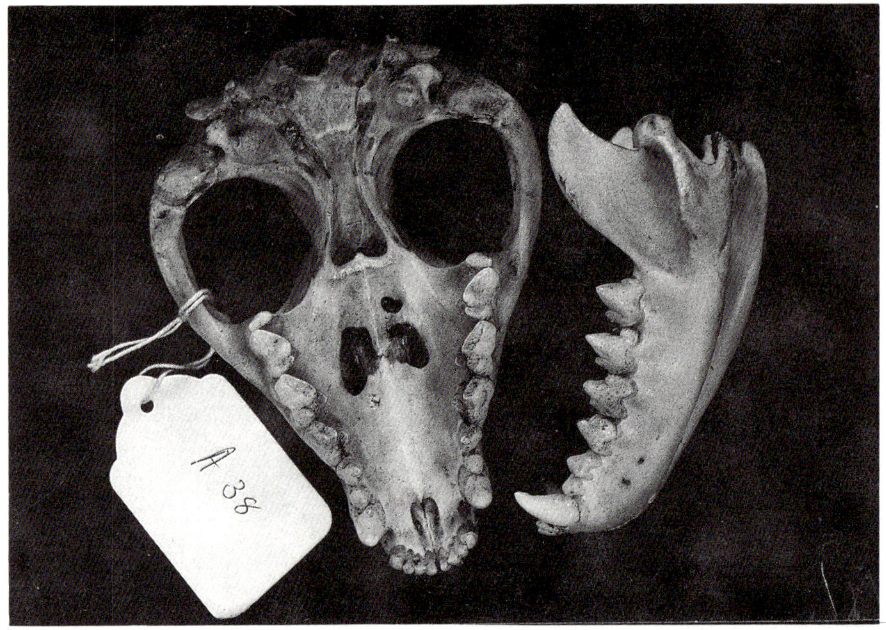

42 The powerful jaws and teeth of the devil are extremely dangerous and the animal needs to be firmly restrained during processing. We placed them in a sack for this where we measured it and made a variety of observations including taking its temperature and a blood sample.

As time went by we made several modifications to our traps. At one stage we added a galvanised sheet iron floor. This was not a good idea as water, urine, faeces and bits of rotten meat collected on the floor. The devils paddled about in this rather rich soup so that handling them was a stinking nightmare. Devils don't smell too good at the best of times but under these conditions they were very hard to tolerate. The trap had to be strong enough to hold the animal but light and small enough to be carried through the scrub.

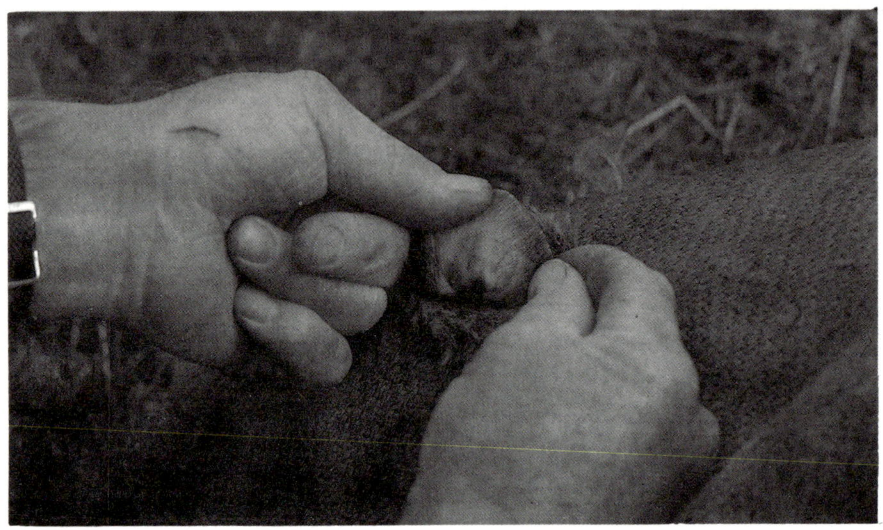

43 We marked devils for future identification by tattooing the left ear. The ear was pulled through a slit in the sack and a number was punched into the ear using a veterinary tattooing kit and Indian ink as a stain. We put the appropriate number into the tattooing clamps, smeared it with the ink and then pressed the number through the ear. These marks lasted the lifetime of the animal.

The devils suffered no serious injuries in the traps. Noses were skinned trying to break through the wire and the tail sometimes was nipped when the trap door closed on it. We found meat baits were best, preferably with a strong, even stinking rotten, smell. Bullock livers were good and they were cheap and went bad quickly but mutton was not favoured by devils.

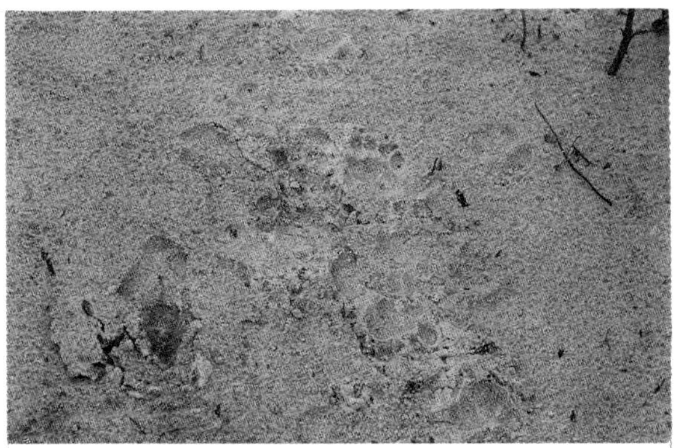

44 It is not difficult to decide if there are sufficient devils in an area to set up a worthwhile trapping program. Their droppings are easily identified and they leave a characteristic footprint. These tracks were at Cape Portland where there were so many that it was impossible to track any individual animal.

45 Devil tracks at Studland Bay. Devils trot in a rocking horse motion with the two hind feet landing almost side by side with the fore feet leaving marks well separated from each other giving a 1:2:1 arrangement. The hind feet of devils are unique among Tasmanian mammals in that they leave a rectangular pad mark.

46 We were able to do a limited amount of trapping at inaccessible places such as the Mainwaring River and Olga River (shown). There were few devils at the Olga as the sedgelands and thick, wet forests supported few animals while at the Mainwaring River the surrounding coastal scrub favoured by wallabies and other marsupials also attracted considerable numbers of devils. Devils live in close contact with humans, especially around farms, even hiding under outbuildings until hunted out by dogs. They frequent the outskirts of cities, even coming into the suburbs.

47 We used a Landcruiser to carry traps, cramming 20 in and around them. Tas Sward is busy getting more traps from our trailer at camp. He was my assistant throughout all the work on the devil. He was an excellent bushman and always cheerful and helpful and I am most grateful to him for his help and interest in the devil research project.

It requires little skill to catch a devil but it needs enthusiasm on a cold and wet day to thrust a hand into a soaking sack containing a rather enraged devil together with the contents of the bottom of the cage. We were lucky and only suffered two minor bites, both due to carelessness from over confidence. We kept our tetanus injections up to date just in case.

We never wore anything on our hands because gloves that were strong enough to give us protection would have been so unwieldy as to be useless. We got a better grip on a wet, struggling devil with bare hands than by using sodden slippery gloves.

Every now and again we would catch a devil which we had not seen for some time and this always caused some excitement. Most were caught several times and one old scrounger at Cape Portland entered the trap twenty eight times, presumably for the free handout.

48 On one occasion we had to use a 1928 Pontiac, Tas Sward in control. Its cross-country performance was almost as good as the Landcruiser.

After being released some devils ran away, some trotted away, others stalked off and a few even sat and looked at us before reluctantly moving off. One stopped in his tracks and returned to eat a piece of meat lying on the ground beside us. None of them could be said to have been terrified by their experiences.

Occasionally we would find that the number of devils coming into our traps would decrease for no apparent reason. The animals never returned to our trapping areas. Perhaps devils under certain circumstances which we do not understand will go 'walkabout'. Old bushman have said that sometimes there would be few devils around in the bush and then, suddenly, there would be a great many and this would last for only a short time. It was as if a wave of roving devils was passing through.

The cause of death including the mass die-offs, of wild devils has not been determined. We caught very few animals in bad condition and these were never caught a second time. One devil was so crippled that it could not walk away but there was no sign of disease. Perhaps it had severe arthritis. When a devil becomes old and cannot either defend itself or run away it probably suffers the fate of the one which we left under the bush at Mount William.

49 Camping in spring is delightful. We only had one wet day in thirteen trips during one November research season. This camp was at Auburn. There usually were two of us in the field party though at times others came along for special projects such as biochemical sampling. We stayed out in the field for up to two weeks.

50 What is the future of the devil?

Changes to habitat during the past two centuries have seen forest or scrub turned into pasture and the forests have been removed from many places which have led to an increase in the native marsupial populations. Devils have benefitted from this perhaps more from the extra food available than in dead stock and garbage.

There are now more devils than at any time since before white settlement. They have no serious predators other than dogs occasionally. Devils are a totally protected species but too many devils around farms can be a pest, especially at lambing time. Permits are issued to property owners to shoot or poison devils where they can be shown to be causing damage.

At the present time the future of the Tasmanian Devil looks good, its greatest protection being its lack of economic value. The only danger would seem to be a return of the disease which has devastated the population several times in the past. It is just a pity the Tasmanian devil is not a more friendly creature.